Boolean Algebra
Is for Children

Boolean Algebra Is for Children

Julio César Martínez Romero

To order additional copies of this book, contact:
Palibrio
1663 Liberty Drive
Suite 200
Bloomington, IN 47403
Toll Free from the U.S.A 877.407.5847
Toll Free from Mexico 01.800.288.2243
Toll Free from Spain 900.866.949
From other International locations +1.812.671.9757
Fax: 01.812.355.1576
orders@palibrio.com
756218

Contents

Acknowledgements

I thank Daniela Carrera Millán and Natalia Pereyra Millán for their kind participation in this book. I also thank and acknowledge the work of Ismael Álvarez León, who kindly took and donated the cover photograph.

I

The Gatekeepers

It is the duty of the gatekeepers to transmit a message according to the rules of their clan. There are five clans of gatekeepers, each with a special code to transmit messages.

A message may be a 0 or a 1.

0 means "don't do anything." It also means "no".

1 means "yes, proceed", "do as instructed", "fulfill your mission", "follow your commands". It also means "yes".

When something happens, it is a "(1)". If it does not happen, it is a "(0)".

II

The NOT gatekeepers

If you have (1) a lot of homework to do, you should better not (0) go to the cinema with your friends.

homework	cinema
1	0

If you do not (0) have any homework, it would be a great idea to go (1) to the cinema with your friends.

homework	cinema
0	1
1	0

The NOT gatekeepers have to follow these rules:

message that NOT receives	message that NOT transmits
0	1
1	0

Fig. 1

receives	delivers
0	1
1	0

Fig. 2

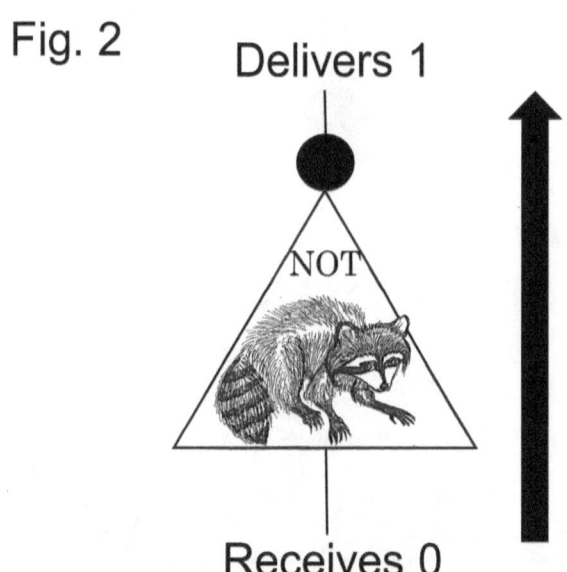

Delivers 1

NOT

Receives 0

Delivers 0

NOT

Receives 1

III

The OR gatekeepers

Maybe today is Naty's birthday, or is it Karina's birthday? If it is neither (0) Naty's, nor (0) Karina's birthday, then there is no need to organize a party.

birthday	party
0, 0	0

If it is (1) Naty's, even if it is not (0) Karina's birthday, we should have a party.

birthday	party
1, 0	1

If it is both Naty's (1) and Karina's (1) birthday, then a party for both will be in order.

birthday	party
0, 0	0
0, 1	1
1, 1	1

These commands were given to the OR gatekeepers:

messages that OR receives	message that OR transmits
0, 0	0
0, 1	1
1, 1	1

Fig. 3

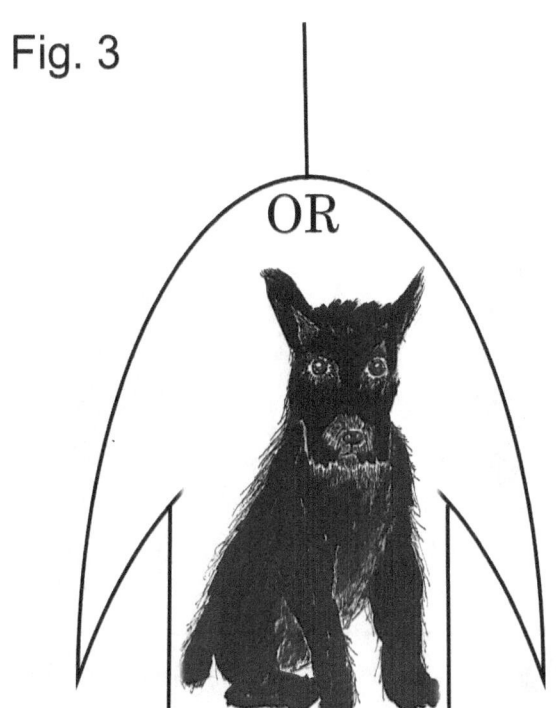

receives	delivers
0 0	0
0 1	1
1 1	1

Fig. 4

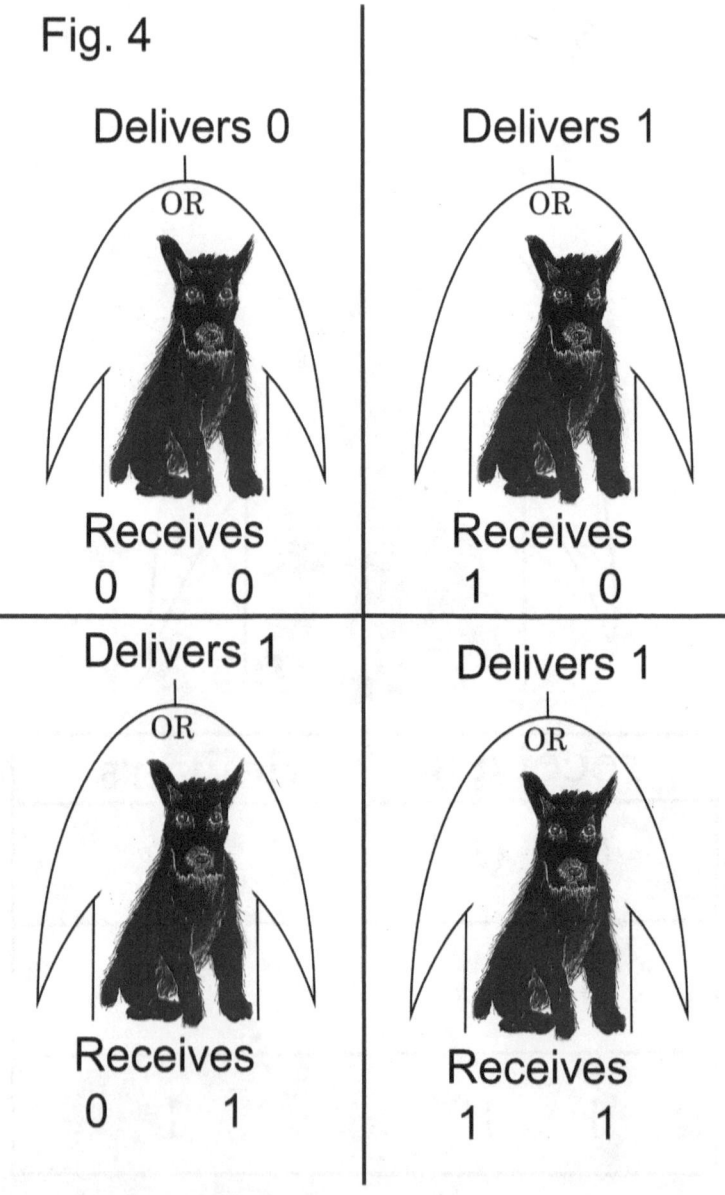

Delivers 0
OR
Receives
0   0

Delivers 1
OR
Receives
1   0

Delivers 1
OR
Receives
0   1

Delivers 1
OR
Receives
1   1

Now, we will present a Boolean circuit and we will find what its final message is. We will follow the circuit step by step.

Fig. 5

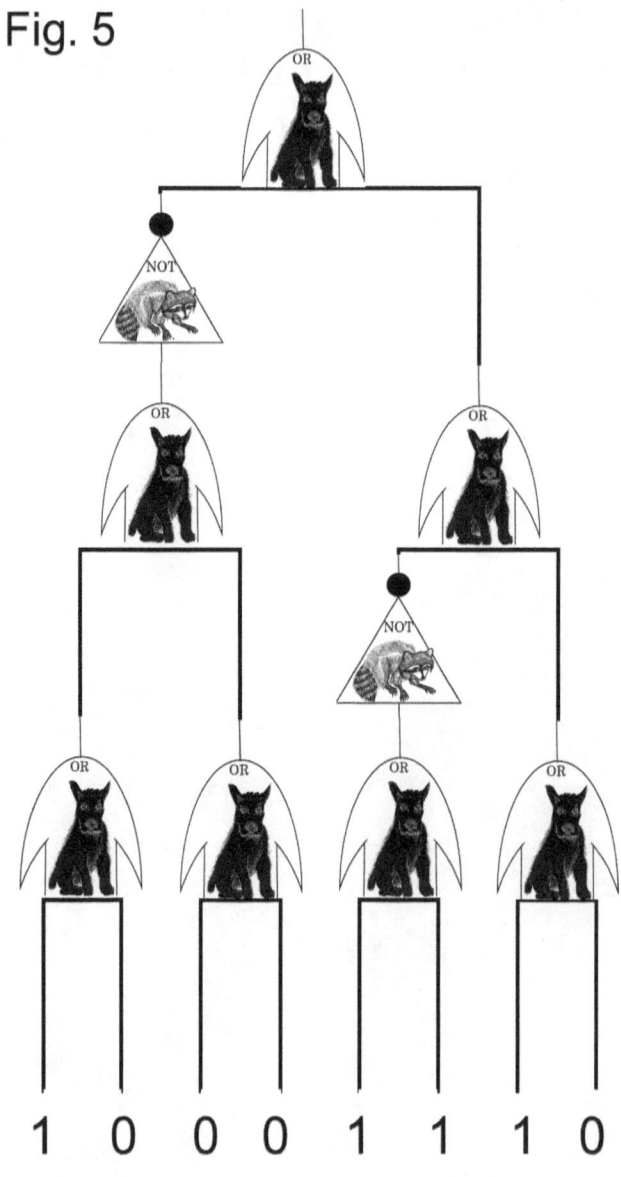

Fig. 6

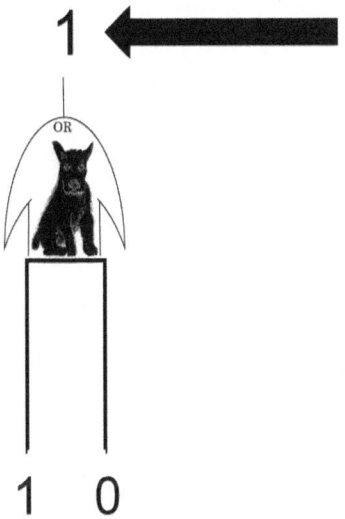

Fig. 7

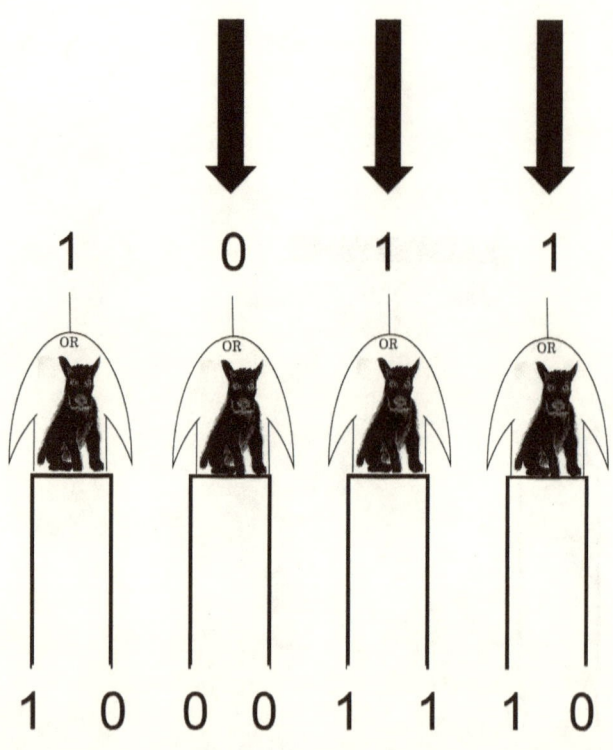

Fig. 8

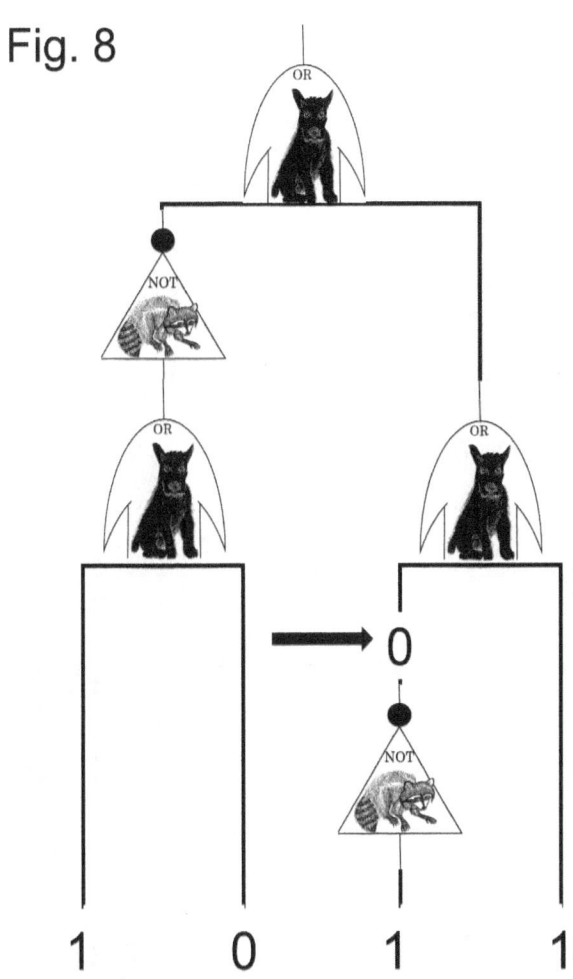

Fig. 9

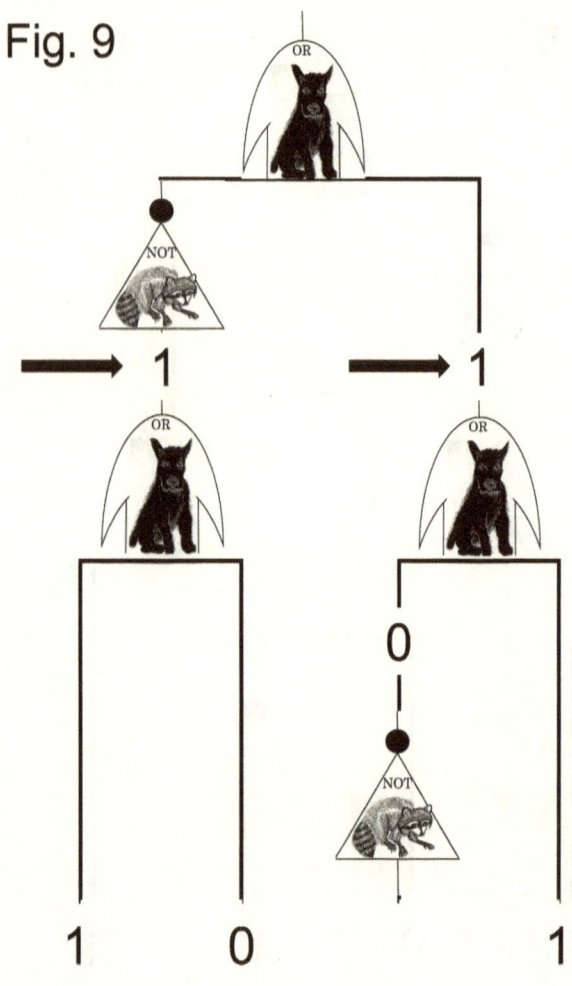

Fig. 10

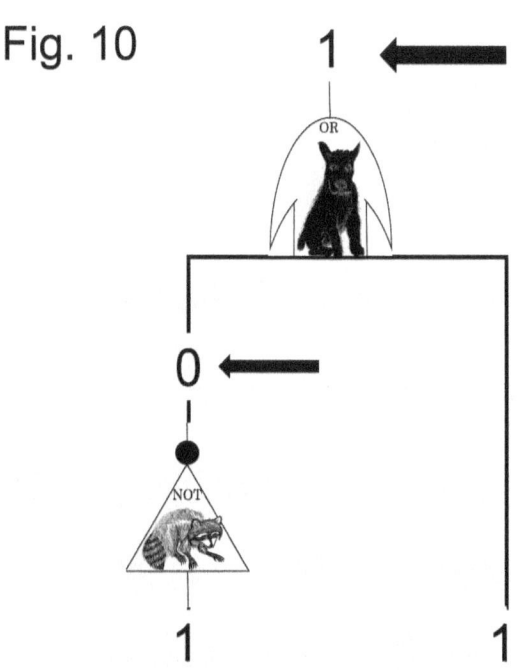

Now, please follow each of the circuits presented, and tell me what its final message is.

Fig. 11

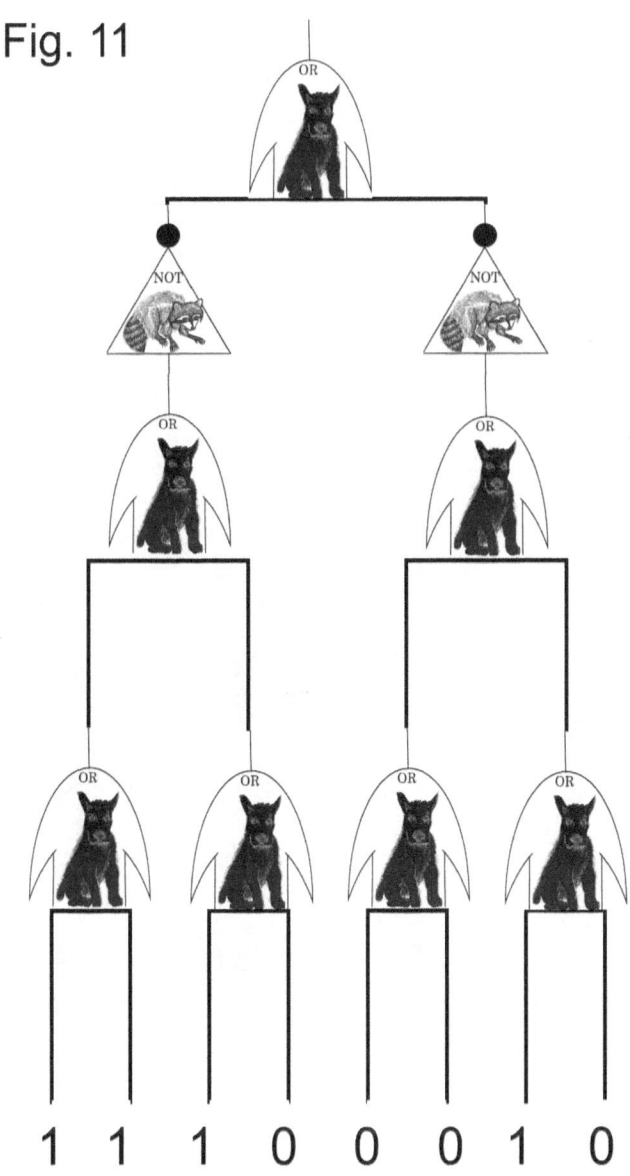

1 1 1 0 0 0 1 0

Fig. 12

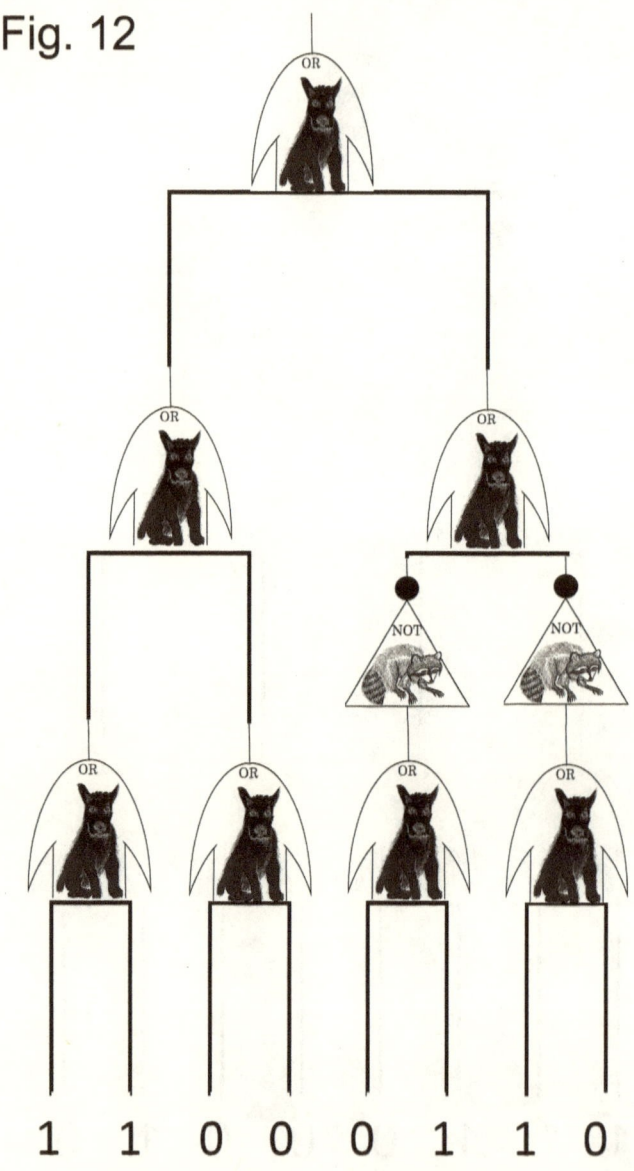

IV

The AND gatekeepers

If Sergio does not (0) bring any beefsteak and Julian does not (0) bring any tortillas, then we will not (0) eat any tacos today.

beefsteak, tortillas	tacos
0, 0	0

If Sergio brings beefsteaks (1) but Julian does not (0) bring tortillas, there will be no tacos (0).

beefsteak, tortillas	tacos
1, 0	0

If Sergio does not bring (0) any beefsteak even though Julian brings (1) tortillas, there will be no tacos.

beefsteak, tortillas	tacos
0, 1	0

Only when both Sergio (1) and Julian (1) bring beefsteak and tortillas, there will be (1) tacos.

beefsteak, tortillas	tacos
1, 1	1

beefsteak, tortillas	tacos
0, 0	0
0, 1	0
1, 0	0
1, 1	1

The AND gatekeepers have been instructed to follow these rules.

AND receives the messages	AND transmits the message
0, 0	0
0, 1	0
1, 0	0
1, 1	1

Fig. 13

receives	delivers
0 0	0
0 1	0
1 1	1

Fig. 14

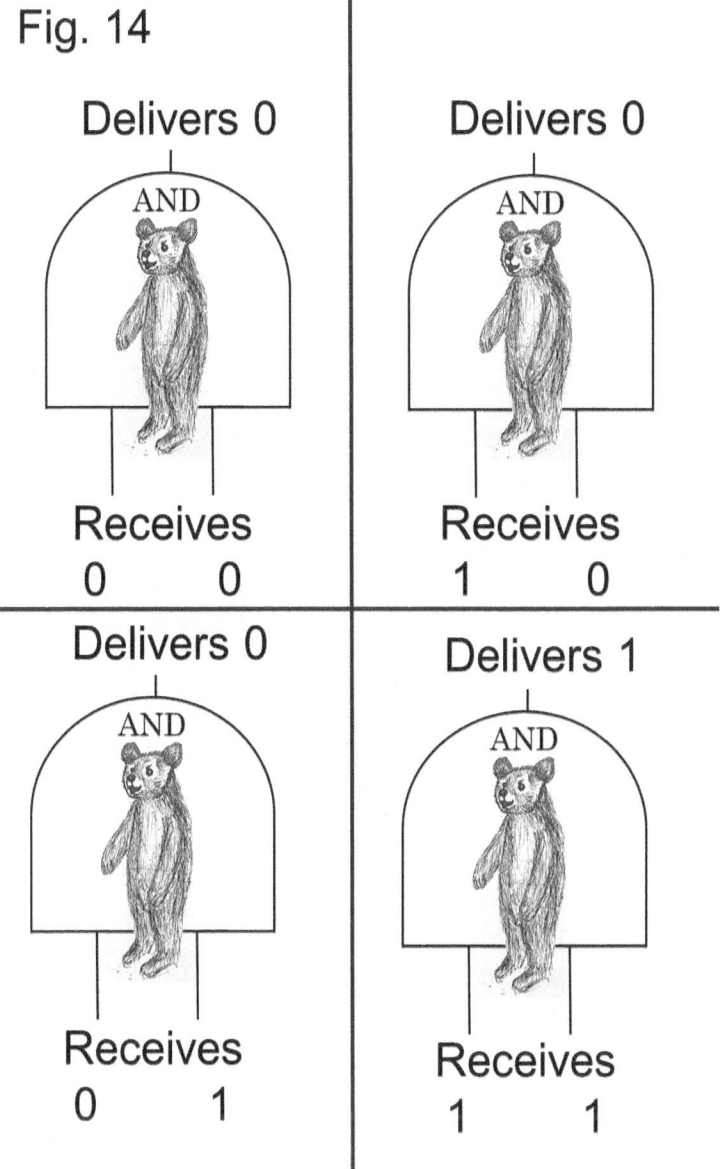

Now, please follow each of the circuits presented, and tell me what its final message is.

Fig. 15

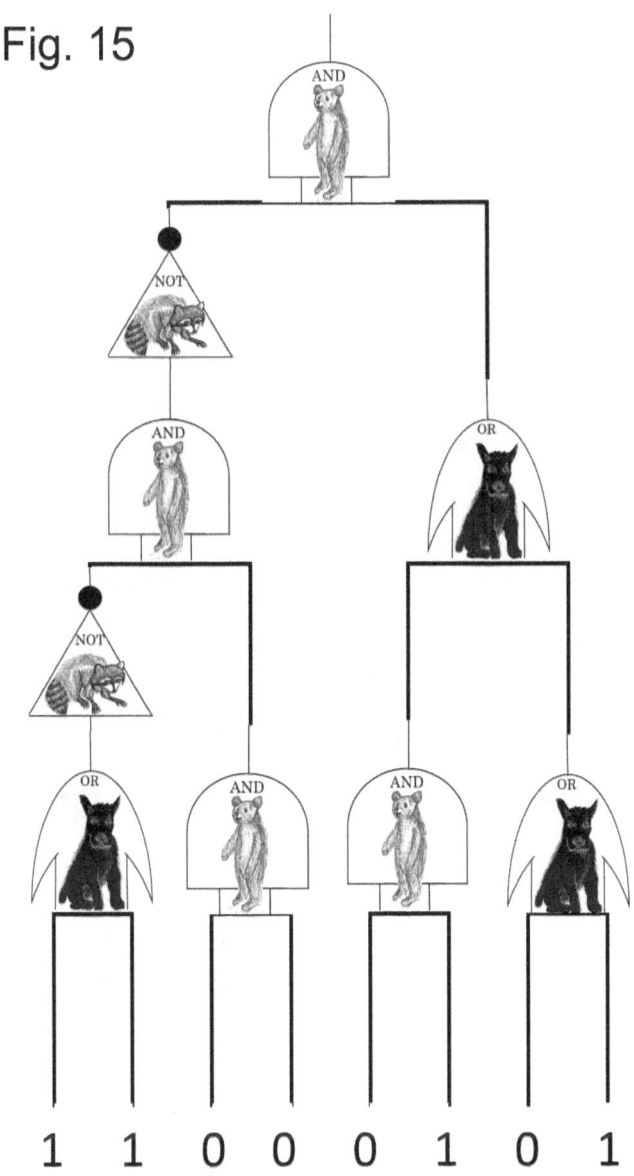

Fig. 16

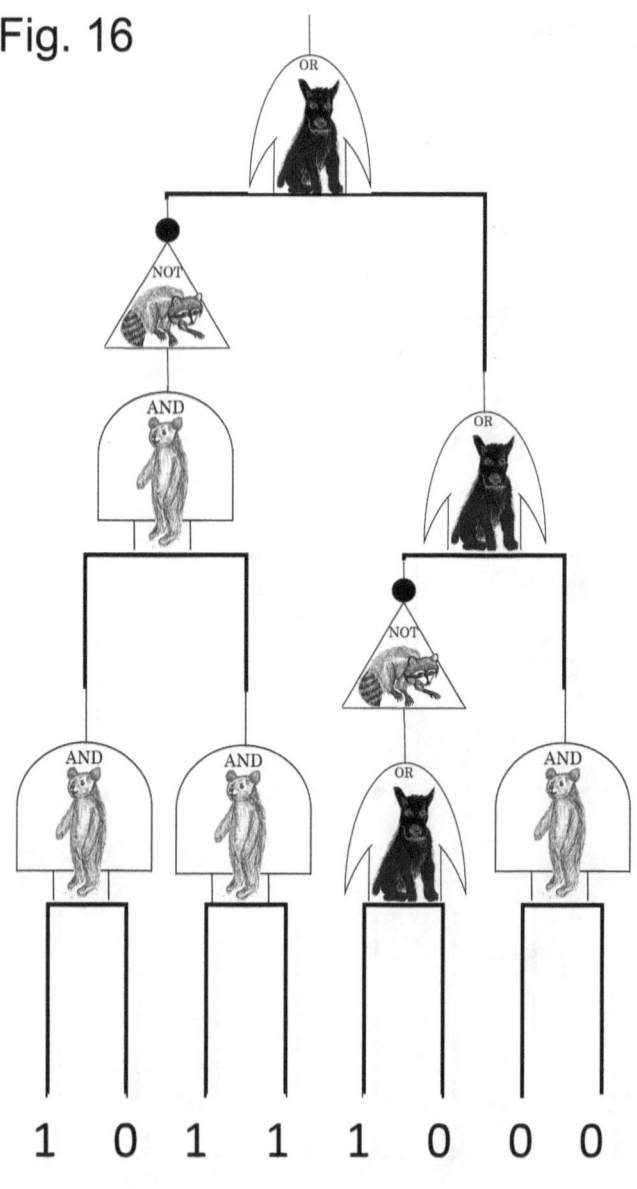

Fig. 17

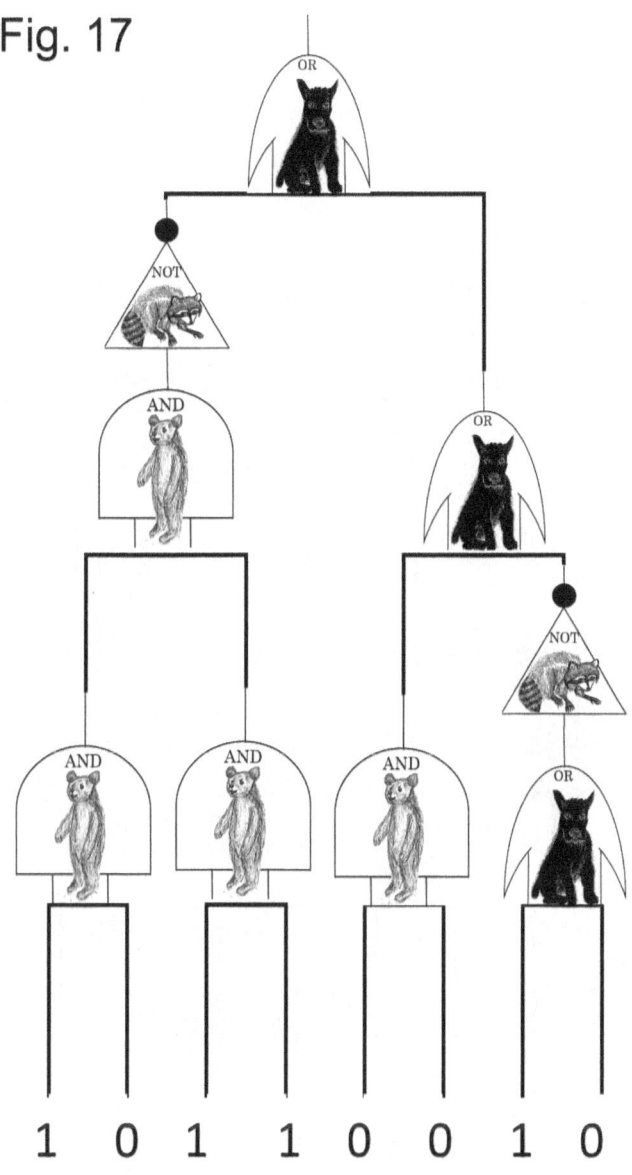

Fig. 18

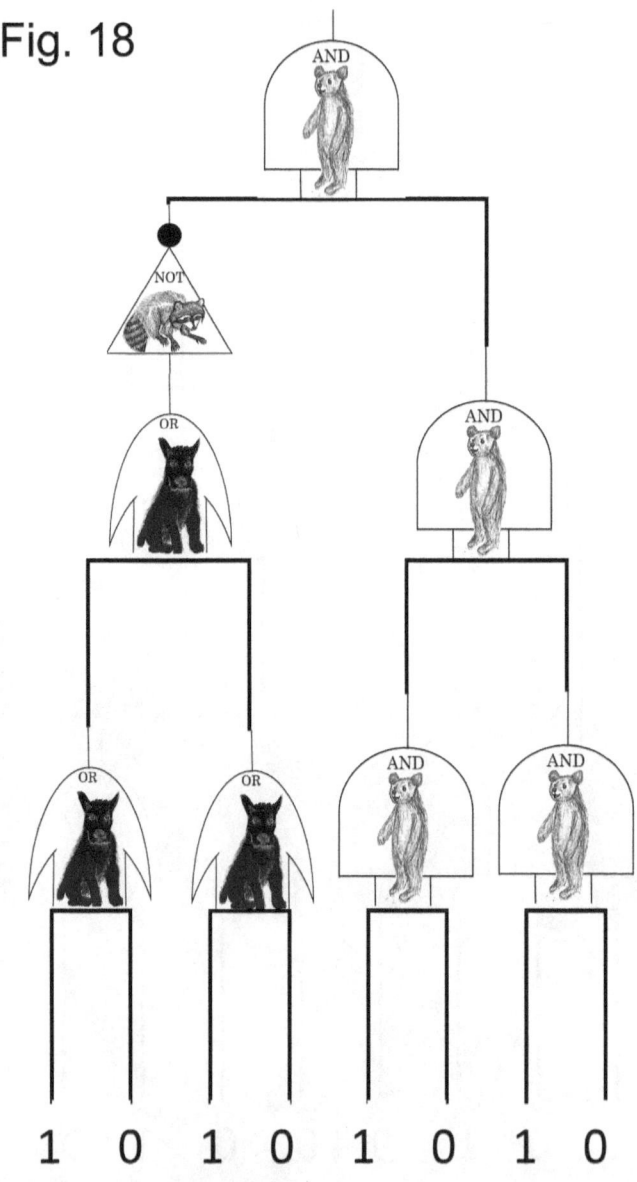

1   0   1   0   1   0   1   0

V

The NOR gatekeepers

Example 1. If the weather is neither (0) rainy nor (0) windy, then we can (1) go to the beach.

rainy, windy	go to the beach
0, 0	1

If the weather is rainy (1), even though it might not (0) be windy, then we do not (0) go to the beach.

rainy, windy	go to the beach
1, 0	0

If the weather is windy (1), even though it might not (0) be rainy, then we do not (0) go to the beach.

rainy, windy	go to the beach
0, 1	0

If the weather is windy (1) and it is also (1) rainy, then we do not (0) go to the beach.

rainy, windy	go to the beach
1, 1	0

rainy, windy	go to the beach
0, 0	1
0, 1	0
1, 0	0
1, 1	0

Example 2. Since Ray and Theresa were forbidden to eat sugar, every birthday party their friends have to be very careful to buy only those cakes that Ray and Theresa dislike. That way neither Ray nor Theresa will be tempted to eat the forbidden dessert.

For example, neither Theresa (0) nor Ray (0) like strawberry cheesecake. Therefore, strawberry cheesecake is a cake that their friends would buy, knowing that Theresa and Ray will not wish to eat it.

Theresa or Ray like a cake	Their friends can buy that cake
0, 0	1
0, 1	0
1, 0	0
1, 1	0

These are the rules for the NOR gatekeeper.

Message that NOR receives	Message delivered by NOR
0, 0	1
0, 1	0
1, 0	0
1, 1	0

Fig. 19

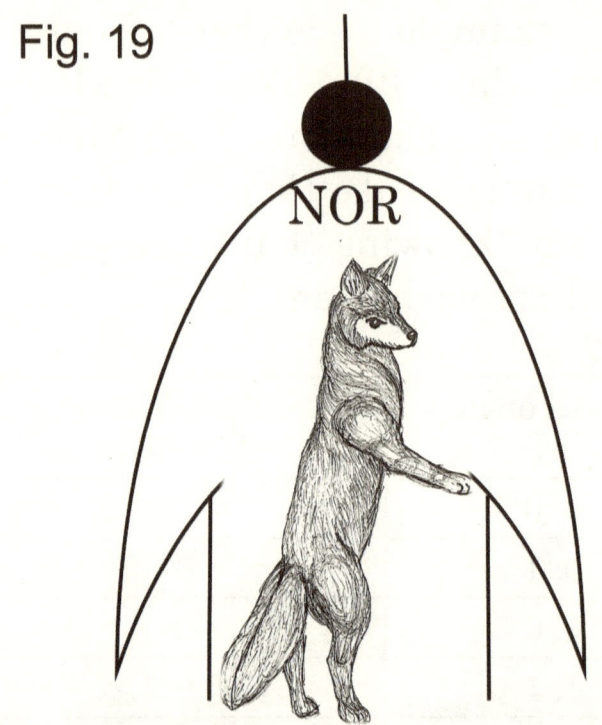

NOR

receives	delivers
0 0	1
0 1	0
1 1	0

Fig. 20

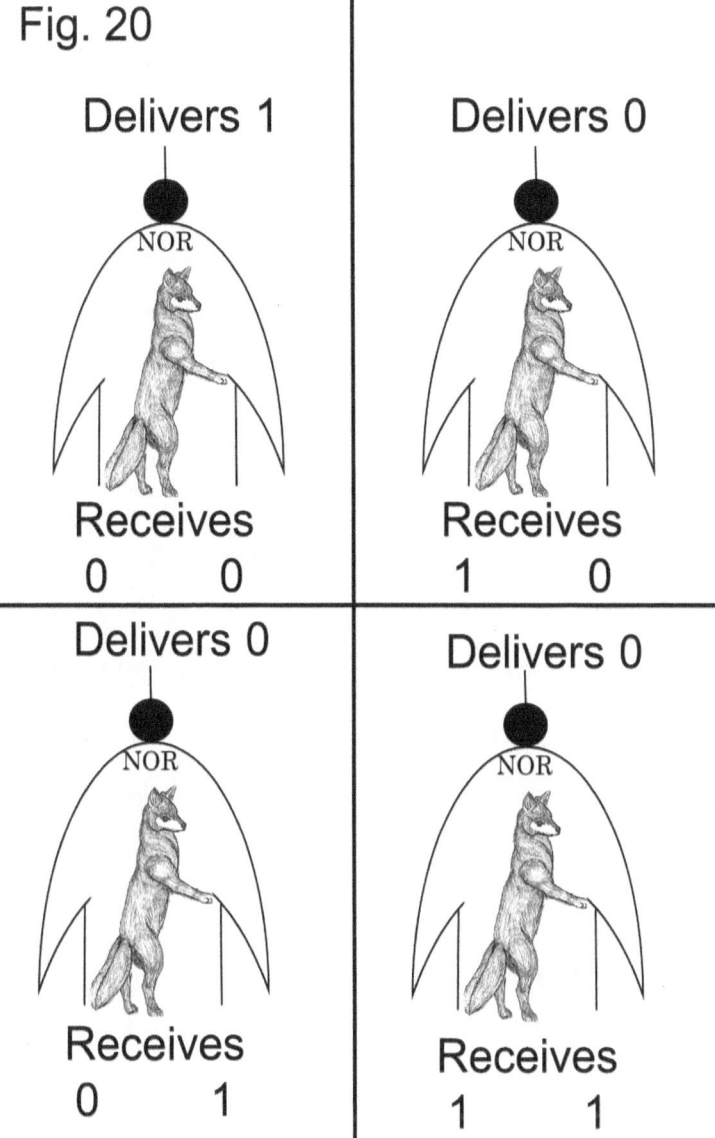

Delivers 1

NOR

Receives
0   0

Delivers 0

NOR

Receives
1   0

Delivers 0

NOR

Receives
0   1

Delivers 0

NOR

Receives
1   1

Now, please follow each of the circuits presented, and tell me what its final message is.

Fig. 21

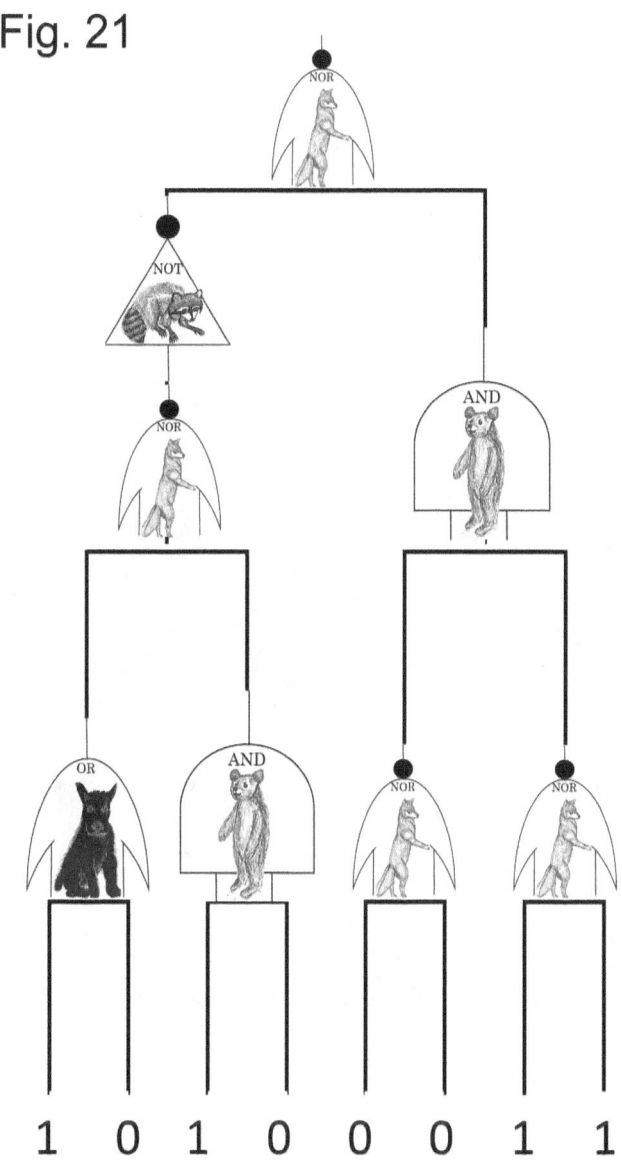

Fig. 22

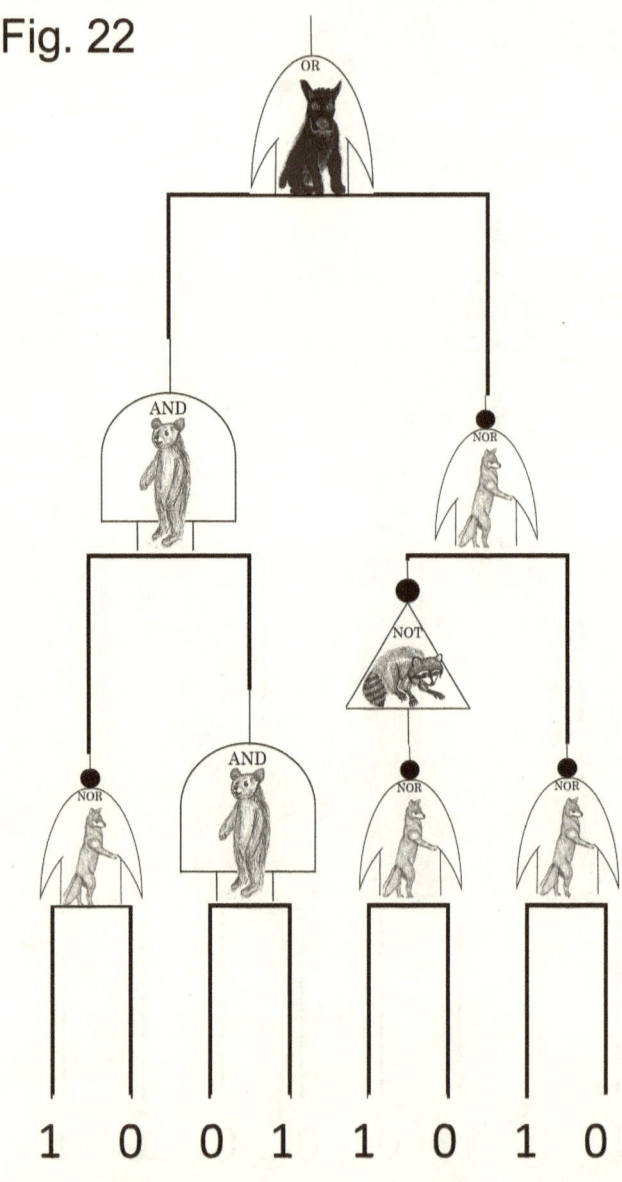

1 0 0 1 1 0 1 0

Fig. 23

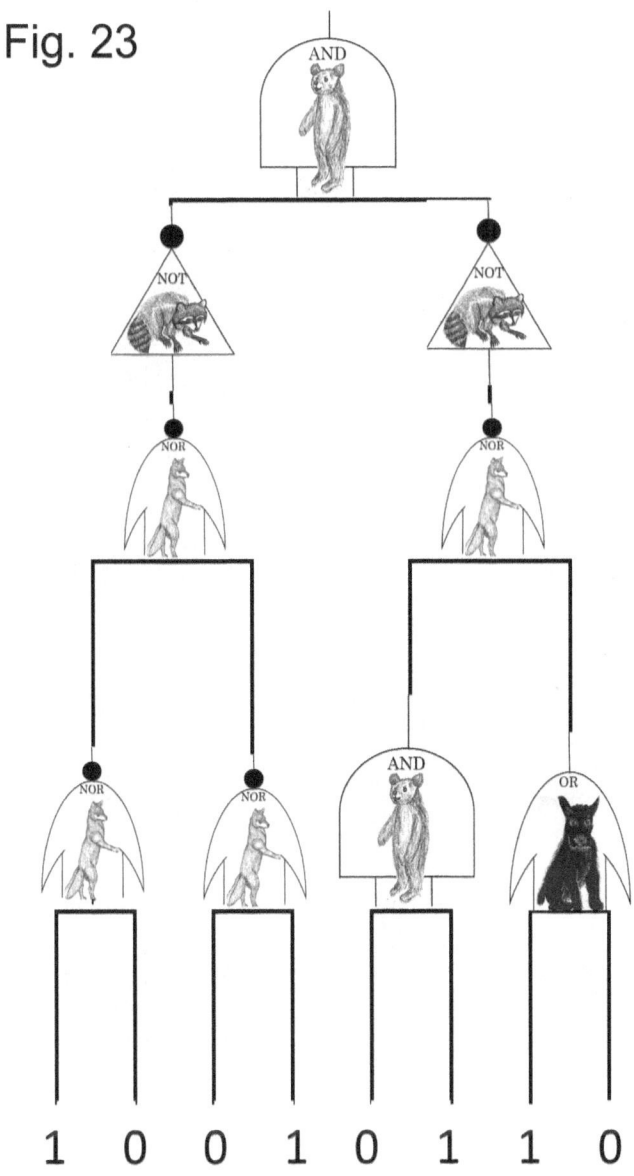

Fig. 24

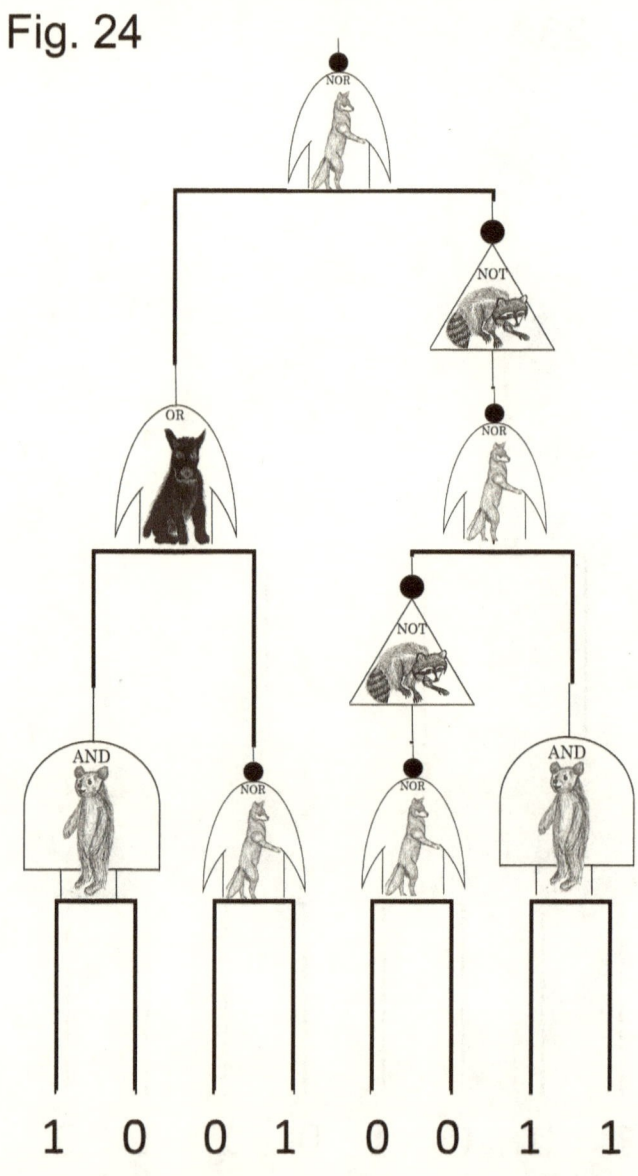

Fig. 25

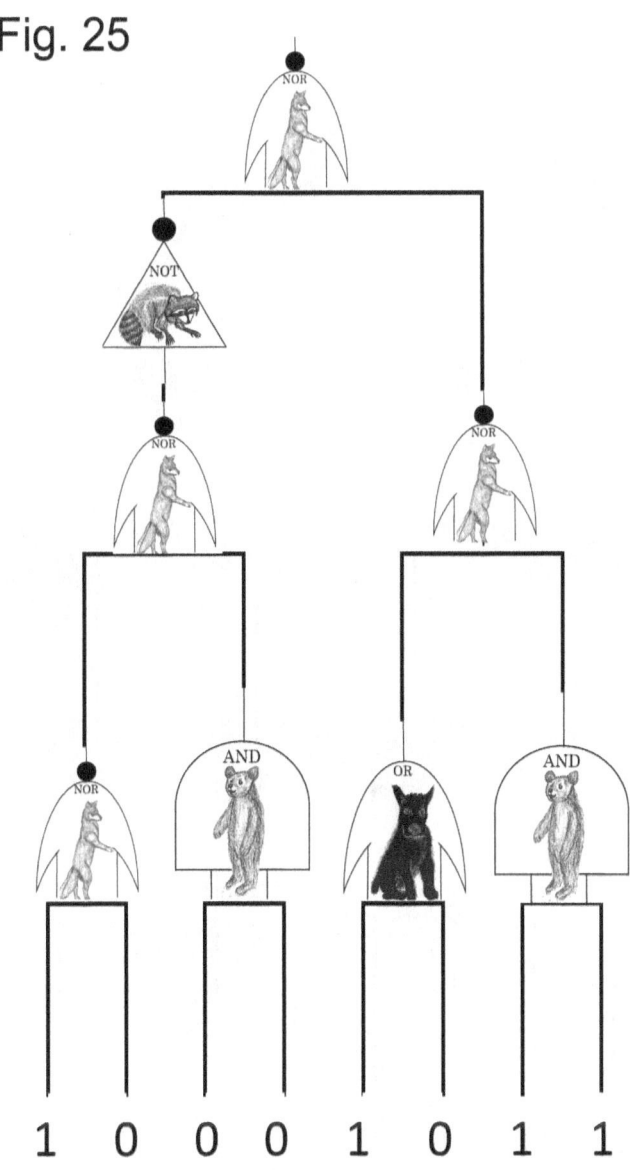

Fig. 26

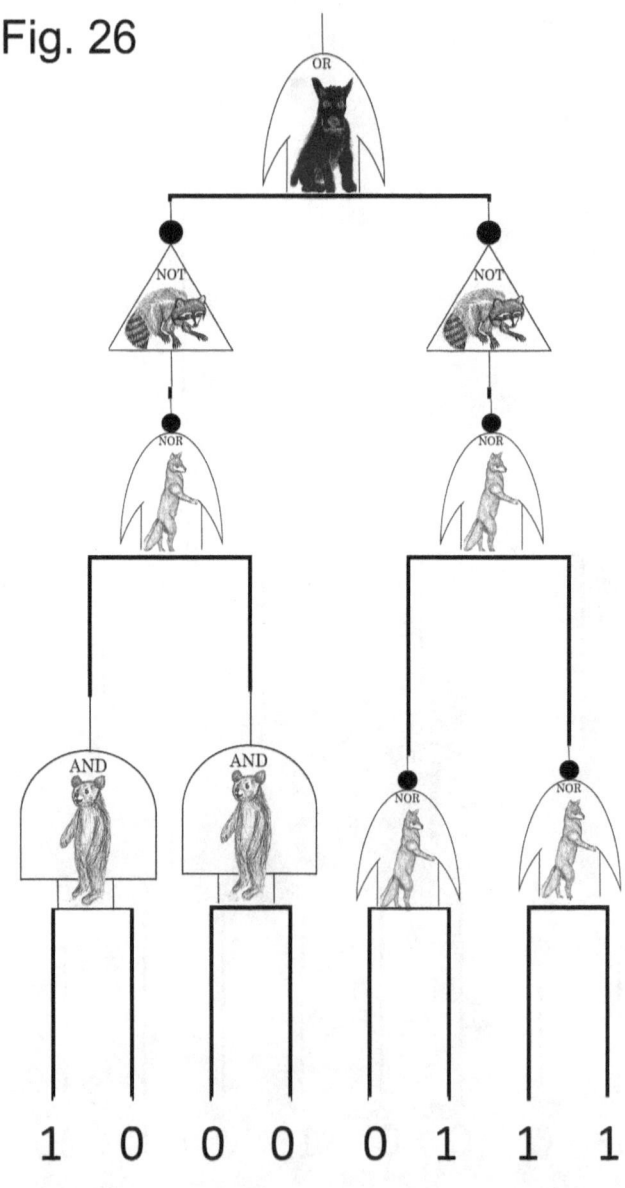

Fig. 27

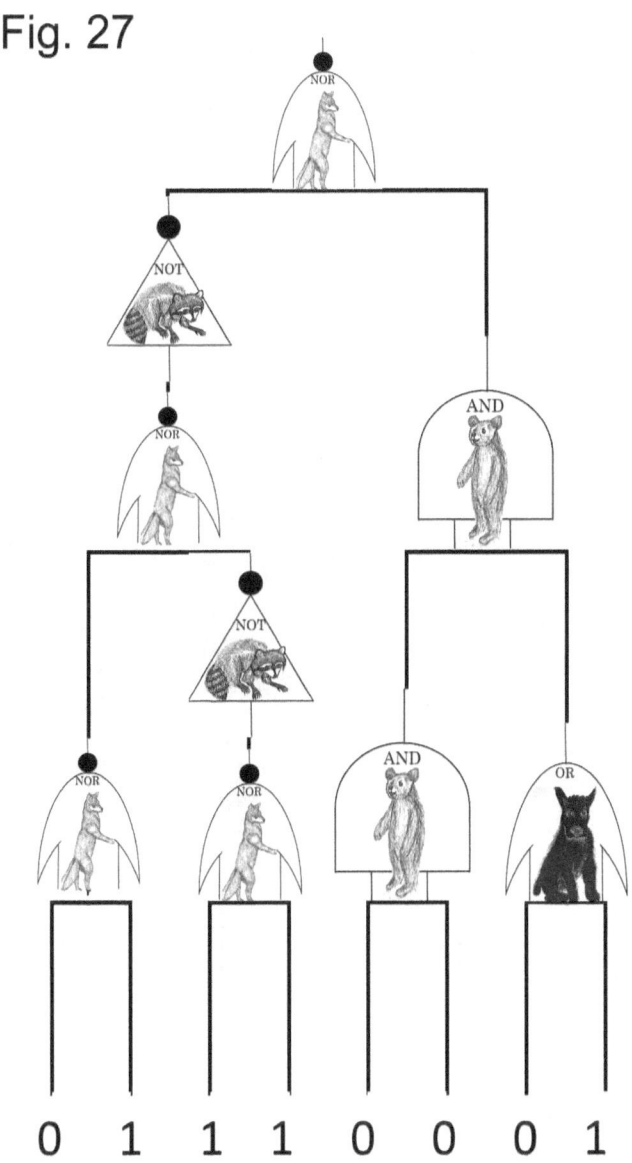

0   1   1   1   0   0   0   1

Fig. 28

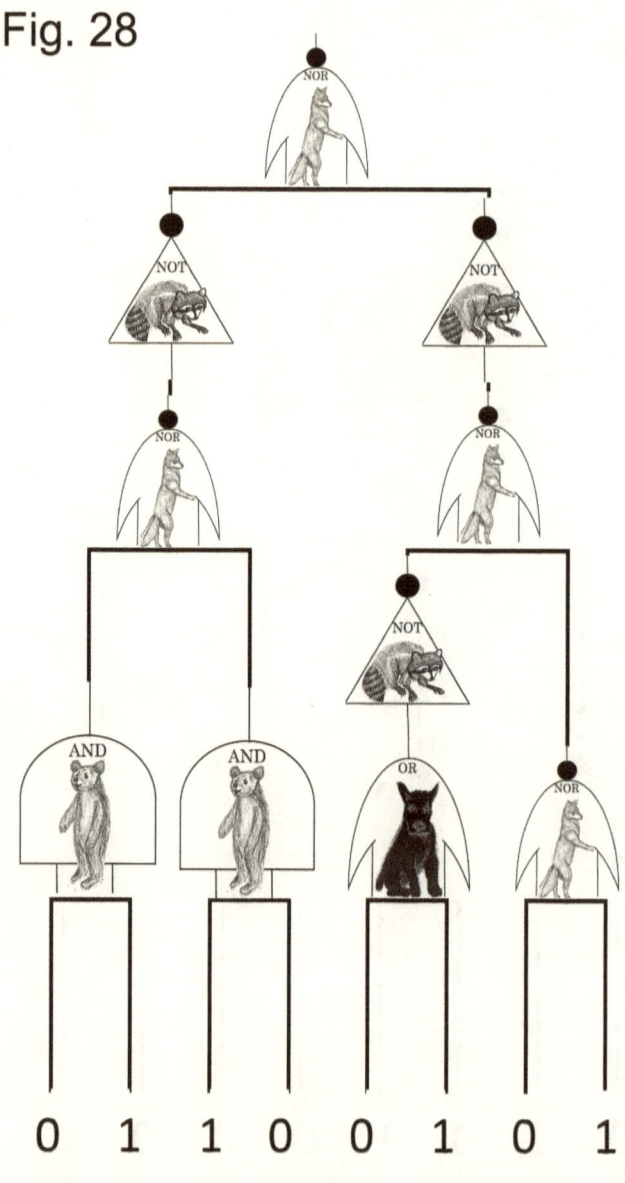

0   1   1   0   0   1   0   1

VI

The NAND gatekeepers

Joe and Charles want to buy a box full of brownies. If neither Joe (0) nor Charles (0) brought any money, then we will have (1) to lend them the money they need to buy the brownies.

Joe, Charles have money	we lend them money
0, 0	1

If Joe does not (0) have any money but Charles has (1) half of the money needed, then we will have to lend them some money.

Joe, Charles have money	we lend them money
0, 1	1

If Joe has (1) half of the money they need but Charles has none (0), then we will have to lend them some money.

Joe, Charles have money	we lend them money
1, 0	1

If both Joe (1) and Charles (1) have enough money to buy the brownies, then we will not have to lend them any money.

Joe, Charles have money	we lend them money
1, 1	0

Joe, Charles have money	we lend them money
0, 0	1
0, 1	1
1, 0	1
1, 1	0

The NAND gatekeepers follow these rules:

NAND receives the messages:	NAND delivers the message:
0, 0	1
0, 1	1
1, 0	1
1, 1	0

Fig. 29

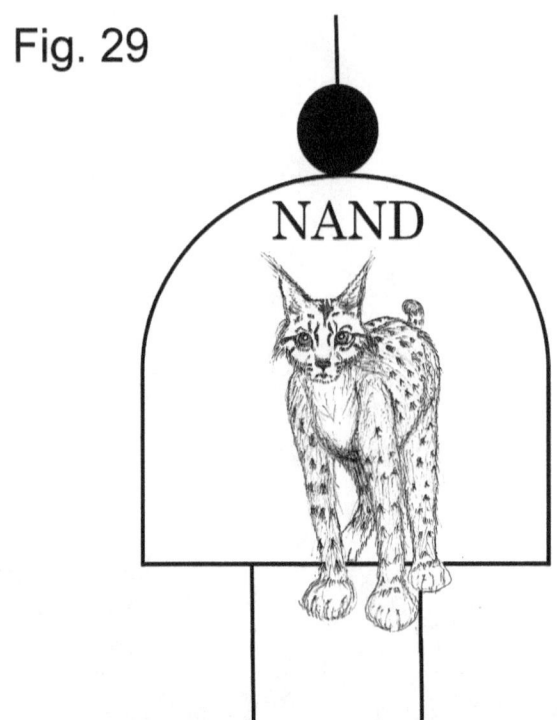

receives	delivers
0 0	1
0 1	1
1 1	0

Fig. 30

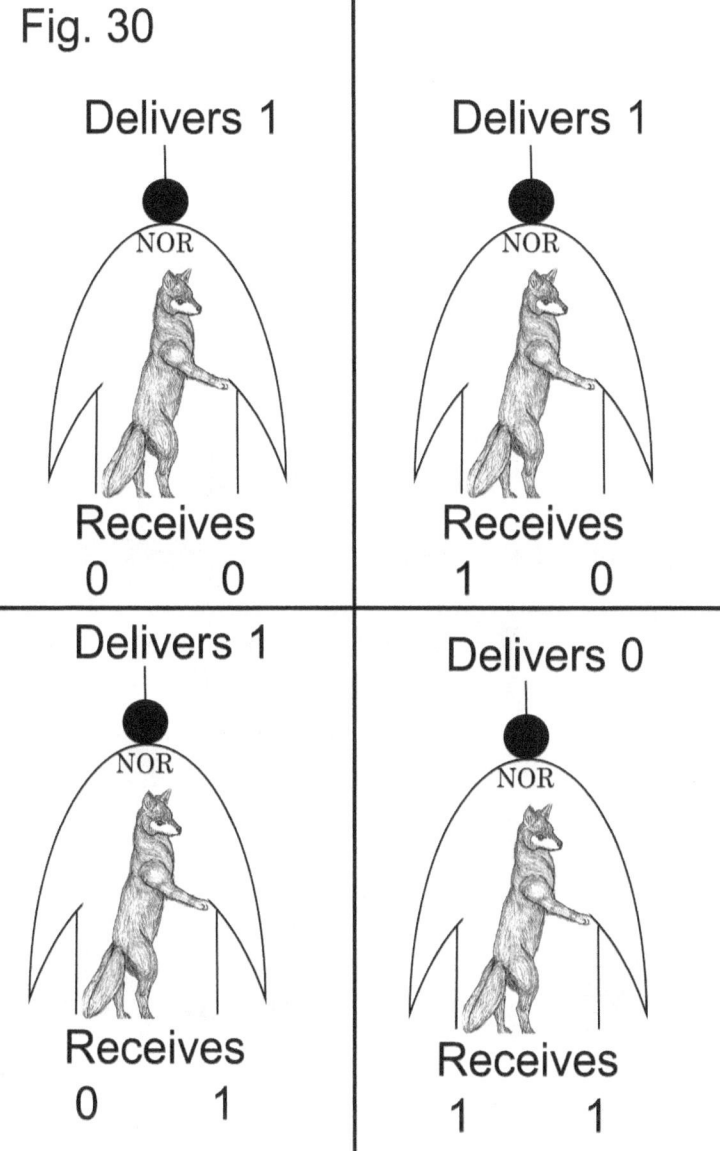

Now, please follow each of the circuits presented, and tell me what its final message is.

Fig. 31

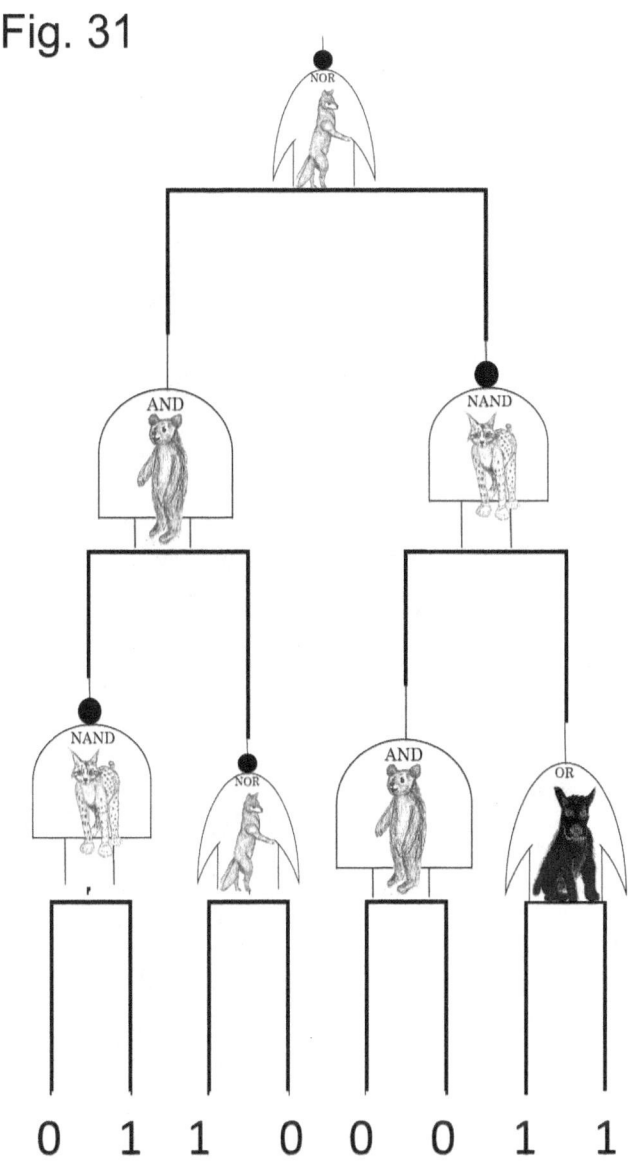

Fig. 32

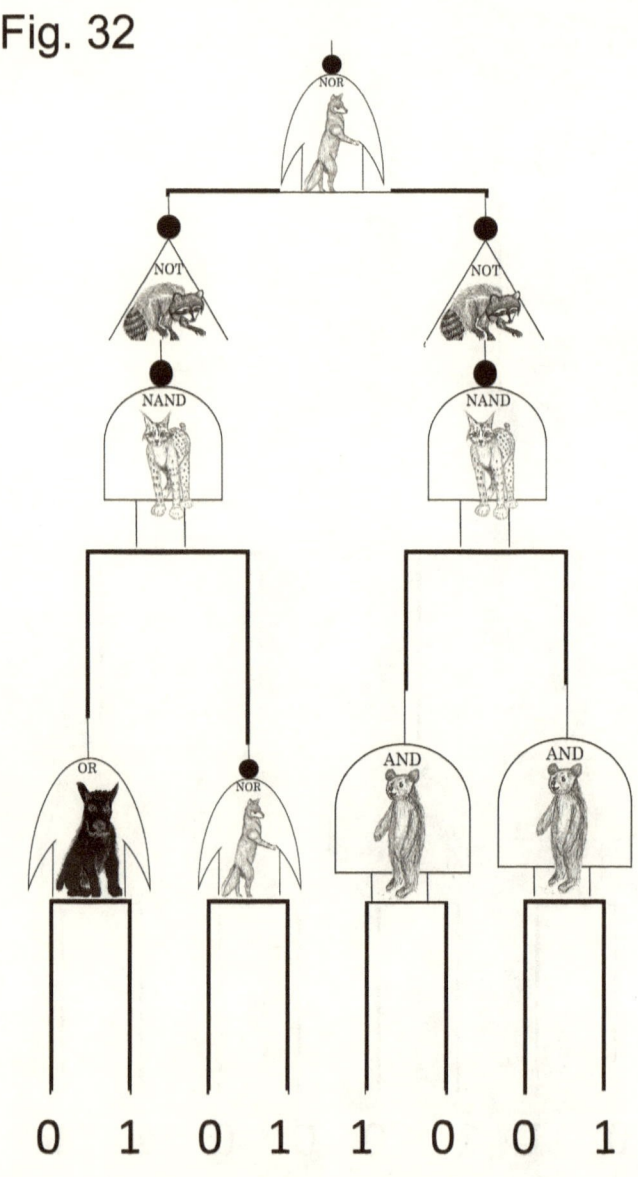

0   1   0   1   1   0   0   1

Fig. 33

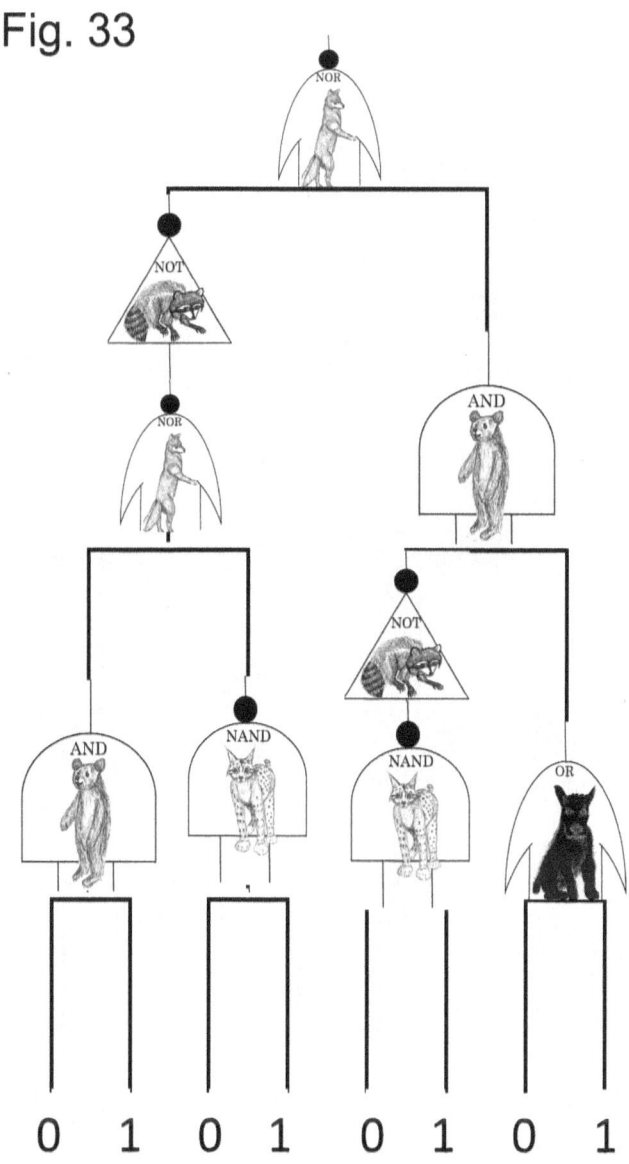

0 1 0 1 0 1 0 1

Fig. 34

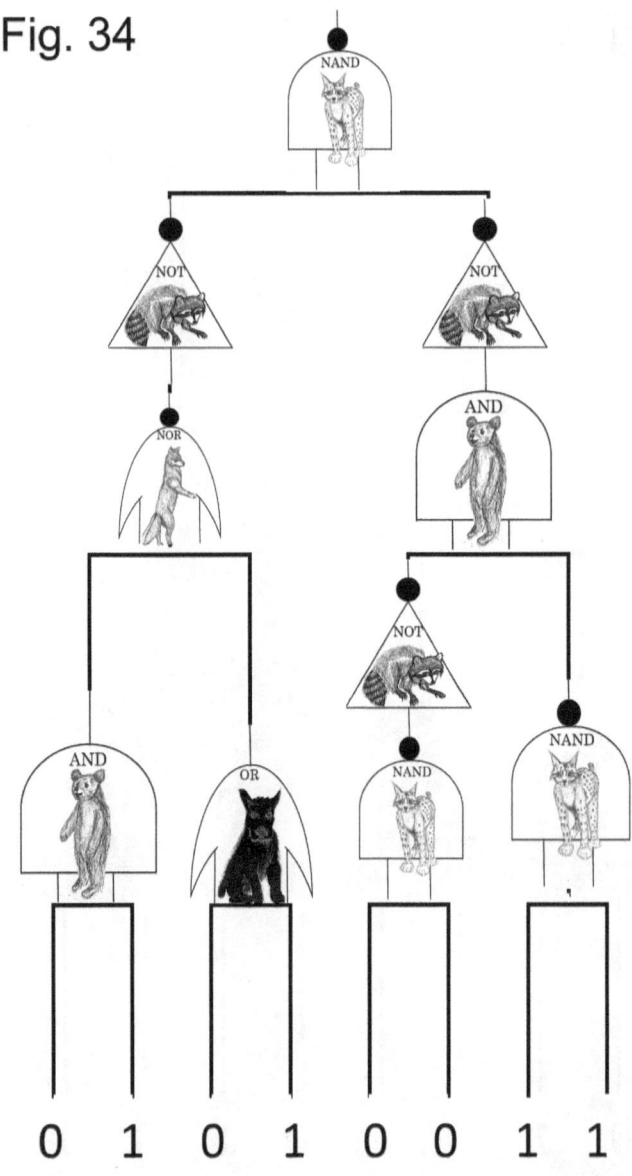

0   1   0   1   0   0   1   1

Fig. 35

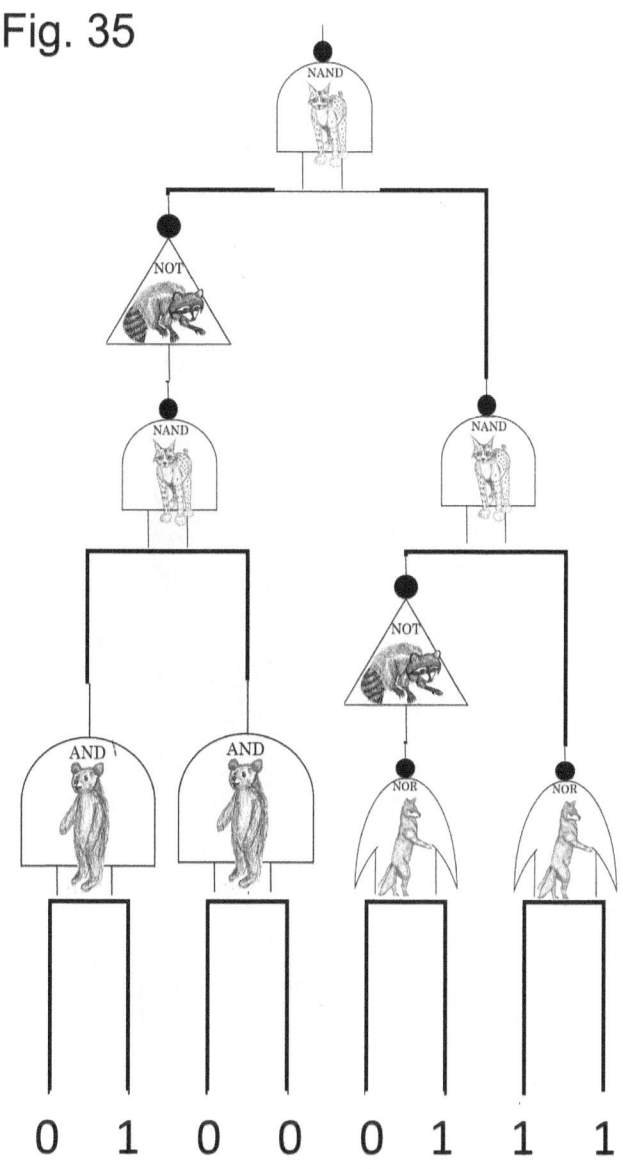

0 1 0 0 0 1 1 1

Fig. 36

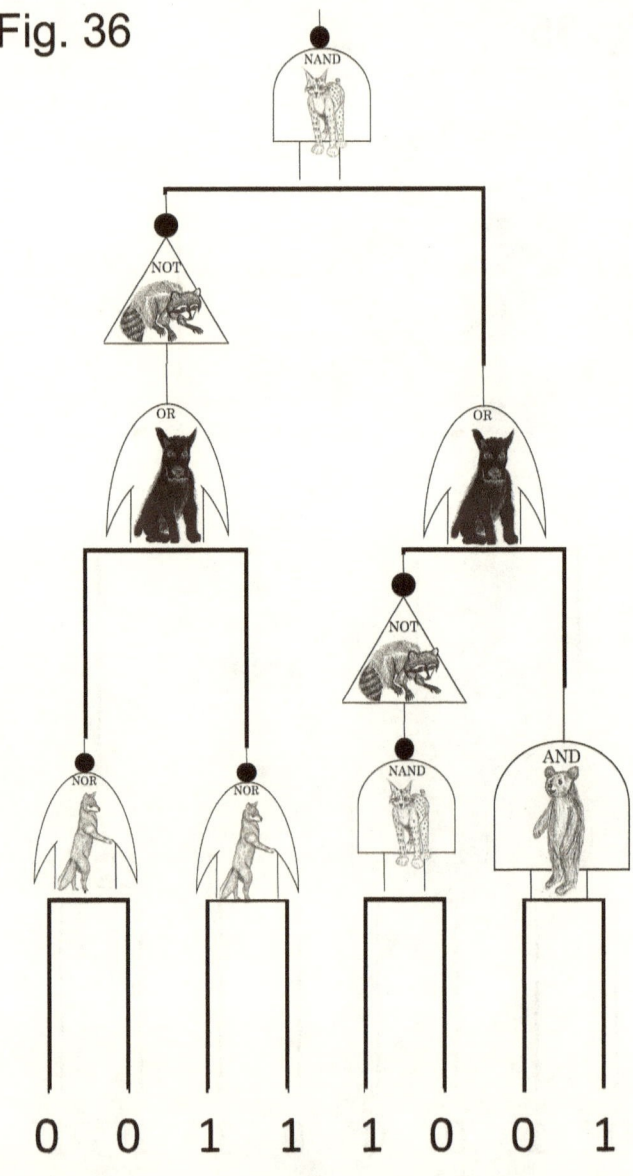

0   0   1   1   1   0   0   1

Fig. 37

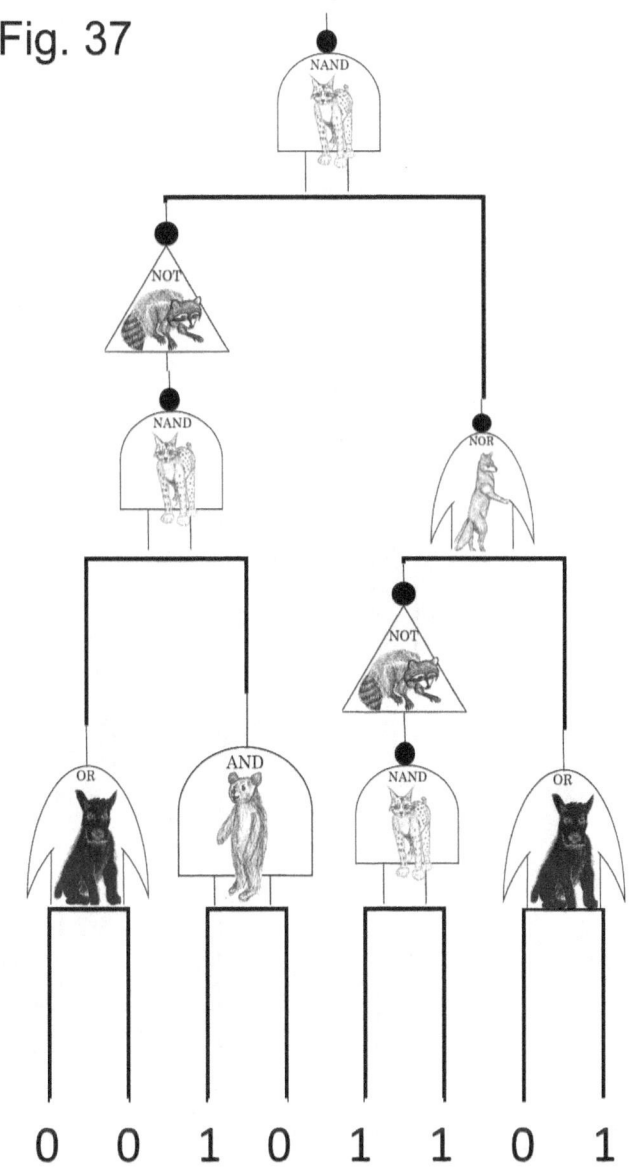

0   0   1   0   1   1   0   1

Fig. 38

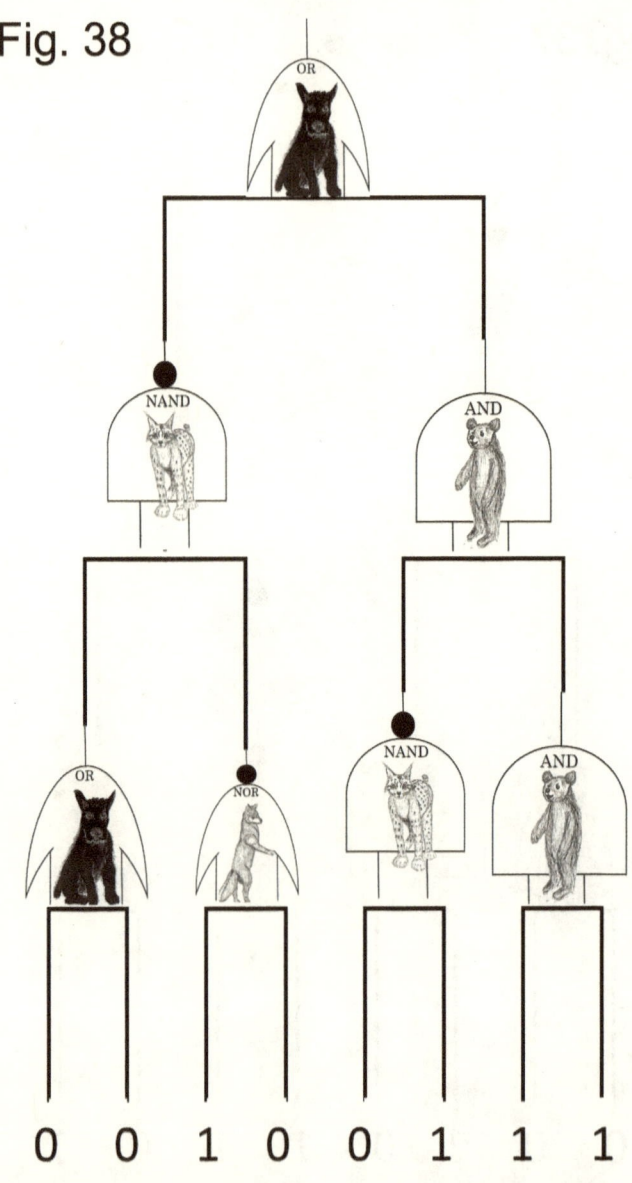

0   0   1   0   0   1   1   1

Fig. 39

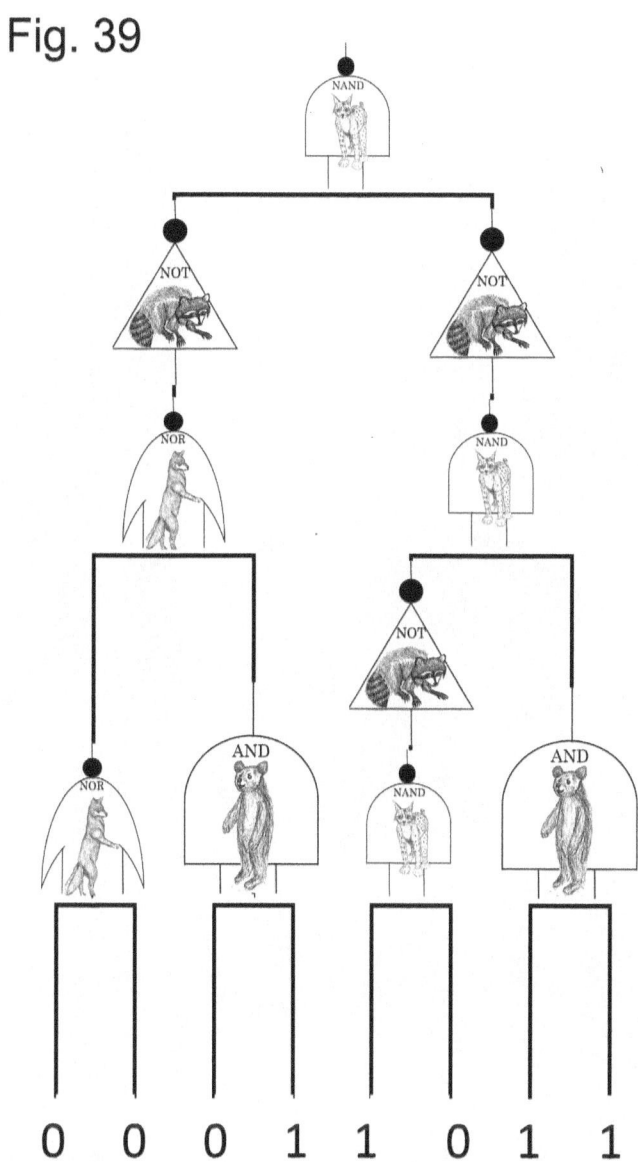

0   0   0   1   1   0   1   1

VII

Hunters and preys

Hunting is the pursuit of meat. At some point in our history, our ancestors became hunters.

Fig. 40

Nevertheless, even as they became hunters, human beings did not stop being the prey of large predators. This posed a dilemma for the hominids. Hunters should be bold, adventurous, and take risks. Preys should be cautious, avoid exposure and minimize their vulnerability.

Fig. 41

On the one hand, the more daring you are, the more you reckless you become. On the other hand, the less risks you take, the less food you will get.

As a result, two types of humans evolved. There are those who are extroverts, who are not afraid to act and there are the introverts, who are very cautious. There is no way to reconcile the two positions. When facing a new experience there are always two contradictory impulses, fear and curiosity. When people are frightened, they miss all opportunities to discover new experiences and enrich their lives.

Fig. 42

Not being afraid leads to risky behaviors, any living being who does not know fear can jump directly into danger and self-destruction.

Fig. 43

Extrovert and shy human beings correspond to two opposite life strategies, boldness and caution. Shy human beings prefer to wait silently, instead of calling attention on themselves. This capability to wait, no matter how long it takes, is typical of prey creatures. A prey animal, that has climbed up a tree, knows that if he comes down, the predator will kill him.

Fig. 44

Therefore, the prey waits as long as it takes until the predator gets bored and walks away. Shy people, like preys, are able to wait quietly and without calling any attention upon themselves. Take for example the behavior of fawns. Newborn fawns are not fast enough to escape a predator. They are reddish in color with small white flecks. This coloration provides camouflage for them to play in the shadows projected by the leaves and branches of the forest. When there is danger, they lie down as close to the ground as they can. Since they lack any odor, if they remain completely immobile, they can go unnoticed. Their secret is to be patient and to wait silently and motionless until the predator is gone.

Fig. 45

VIII

A dear robot deer

A fawn is the cutest creature you will ever see. Anyone would love to have one as a pet. Unfortunately, a fawn will grow up too soon and he will become a destructive and aggressive buck. If you really like wild animals, you should never have one as a pet. Let them be wild. Nevertheless, if you absolutely need to have a deer as a pet, then maybe you should consider building a robot deer. This way, you will have a pet deer, and wild deer will roam free in the forest.

Fig. 46

The first step in building a robot deer is to design the logic circuits that will lead his behavior. You will need to ask yourself a fundamental question. What do you expect from a robot pet?

IX

Cake and ice cream

The first task that any reasonable owner of a robot deer would ask to his/her dear robot is to bring you cake and ice cream when you ask for them.

The cake can be either hazelnut mocha or strawberry shortcake. The ice cream should be vanilla and chocolate.

Your robot deer opens the refrigerator and finds that there is (1) mocha hazelnut cake but no (0) strawberry shortcake. There is (1) also vanilla ice cream but no (0) chocolate ice cream. Will your dear robot be able to perform the task assigned to him?

Let us design the Boolean circuits for your robot.

Fig. 47

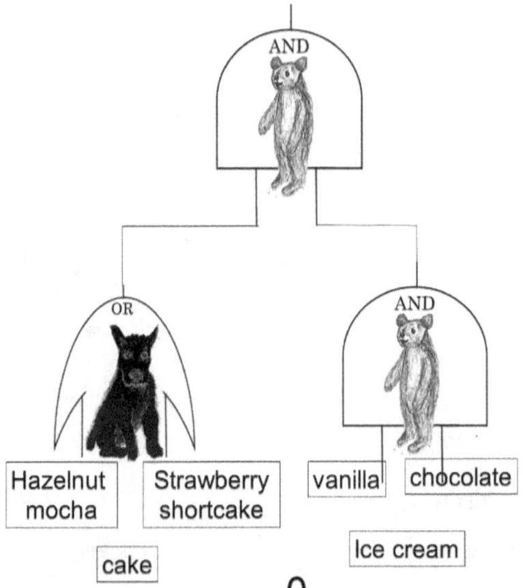
AND
OR
AND
Hazelnut mocha
Strawberry shortcake
vanilla
chocolate
cake
Ice cream

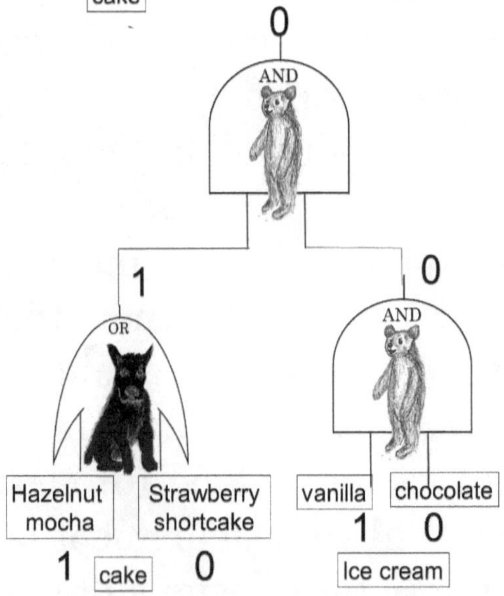
0
AND
1
0
OR
AND
Hazelnut mocha
Strawberry shortcake
vanilla
chocolate
1
0
1
cake
0
Ice cream

X

Bermuda shorts and shirts

Today you will send your robot deer shopping. You were invited to play mini golf and you want to buy new Bermuda shorts and a shirt. The Bermuda shorts should be either blue or green and the shirt should have short sleeves and be made of cotton.

If your robot buys blue Bermuda shorts and a cotton short sleeves shirt, is his performance correct?

Let us design the Boolean circuits for your robot.

Fig. 48

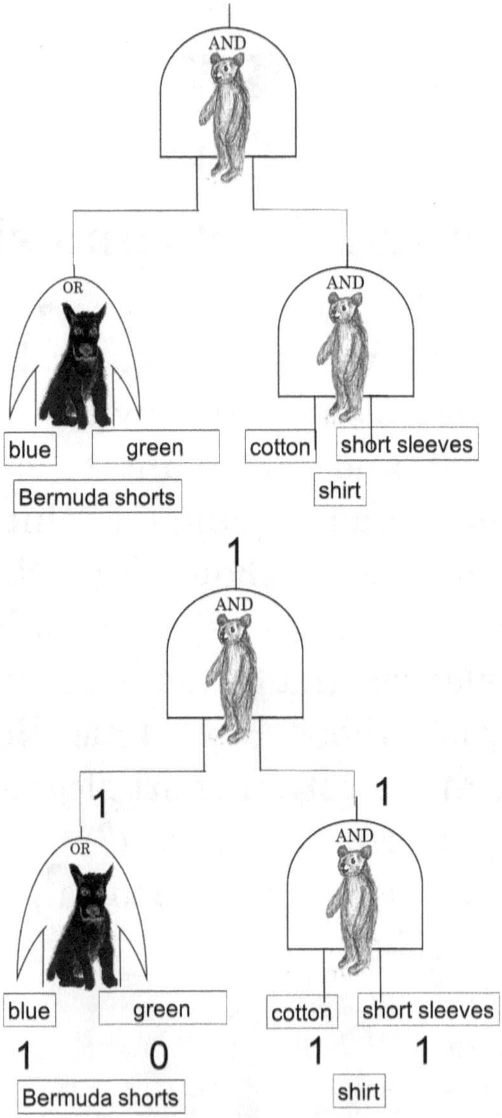
AND
OR
AND
blue
green
cotton
short sleeves
Bermuda shorts
shirt
1
AND
1
1
OR
AND
blue
green
cotton
short sleeves
1
0
1
1
Bermuda shorts
shirt

XI

Films

You invited some friends to watch a film with you at home. Jennifer only likes 3D animations and Regina only likes films with a little romance. William does not like drama and Leonard likes a lot of action. You ask your dear robot to download the film that would be adequate for you and your friends to watch.

The robot deer downloads a 3D animation with lots of action and some romance. In fact, there is almost no argument or drama, just continuous action. Will your friends like the films chosen by your robot?

Let us design the Boolean circuits for your robot.

Fig. 49

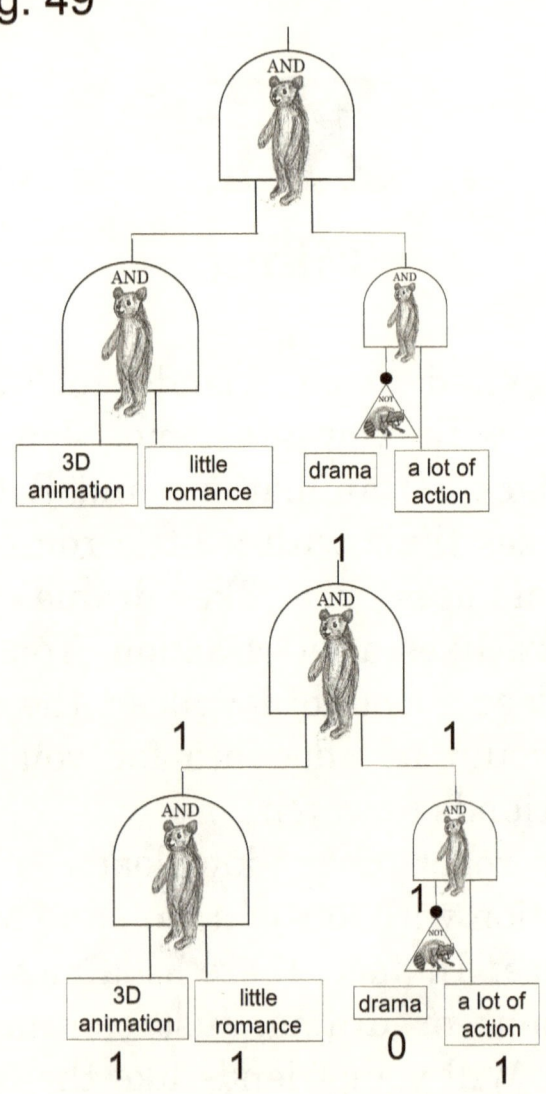

XII

Book report

This weekend you have to write a book report but you still do not know what book to read. You think that you are in a mood for either a historical novel or a science fiction story. The book must not be too long and it should be a real page turner. You ask your dear robot to find the perfect book for you.

Your robot deer brings you and amazing Sci Fi short story full of unpredictable twists and an original and surprising argument. Did your robot choose the right book for you?

Let us design the Boolean circuits for your robot.

Fig. 50

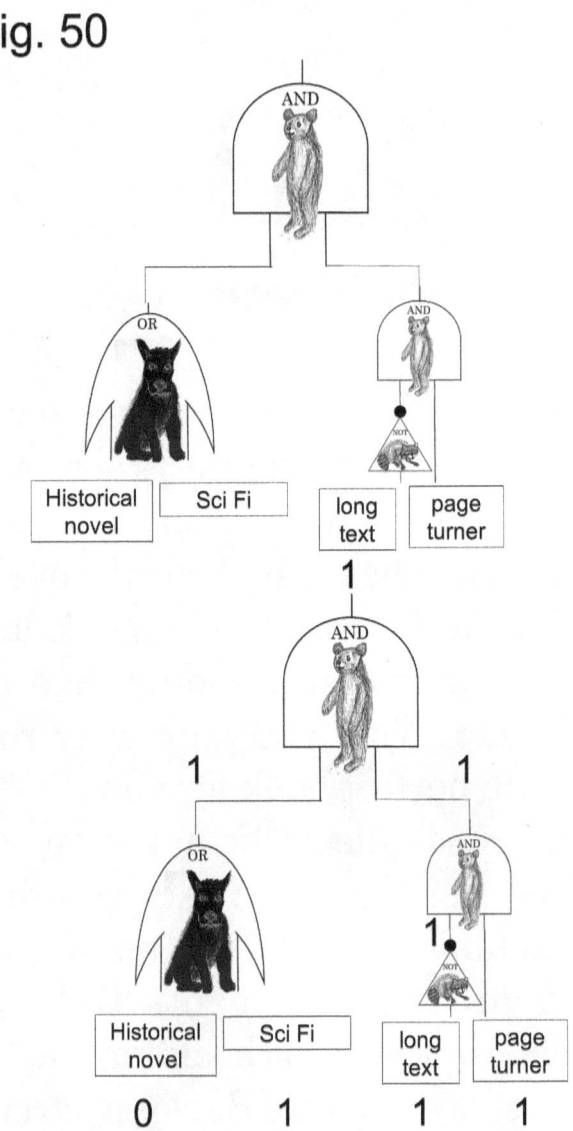

XIII

The While loop

Now you want your robot to be persistent. You want him to try over and over until he gets exactly what you want. Every time that he finds a solution, he must ask himself if the solution fulfills the conditions imposed on him.

This is called the While loop and it is represented as follows:

Fig. 51

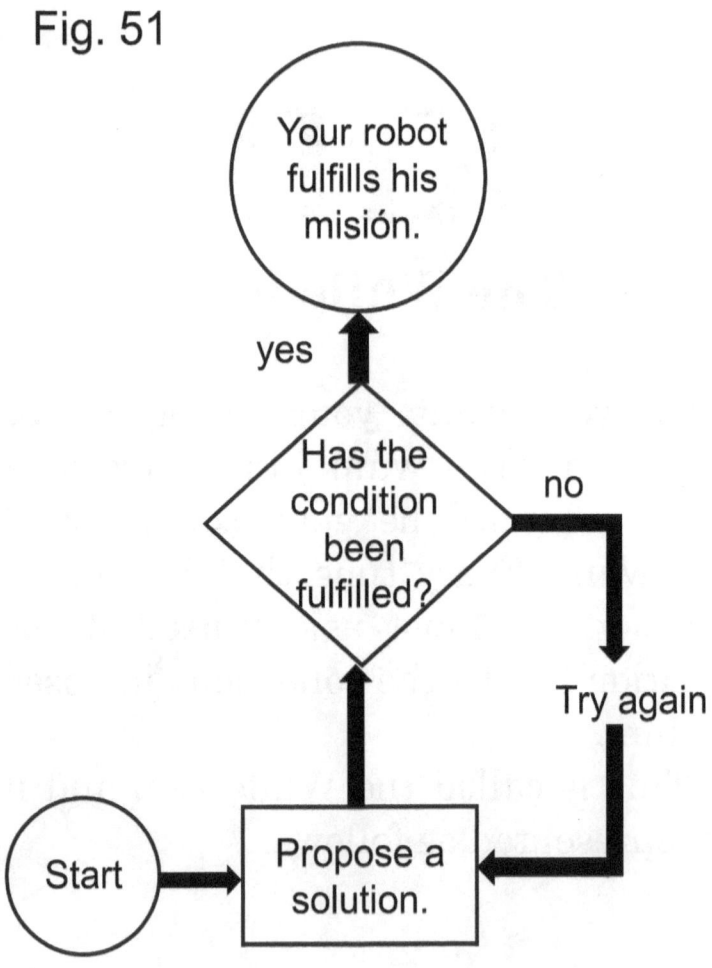

XIV

Italian food

Today you would like to eat either pizza Margherita or fettuccine Alfredo, and either almond biscotti or triple chocolate tiramisu.

Let us design the Boolean circuits so that your robot will know his instructions.

Fig. 52

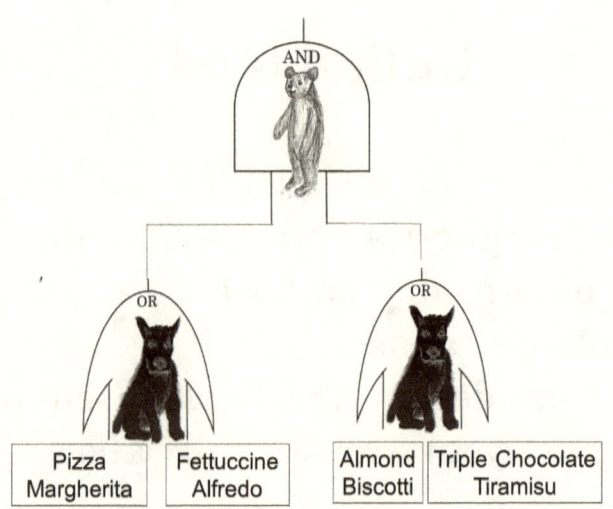

This is the design for the while loop of the instructions that your robot will follow.

Fig. 53

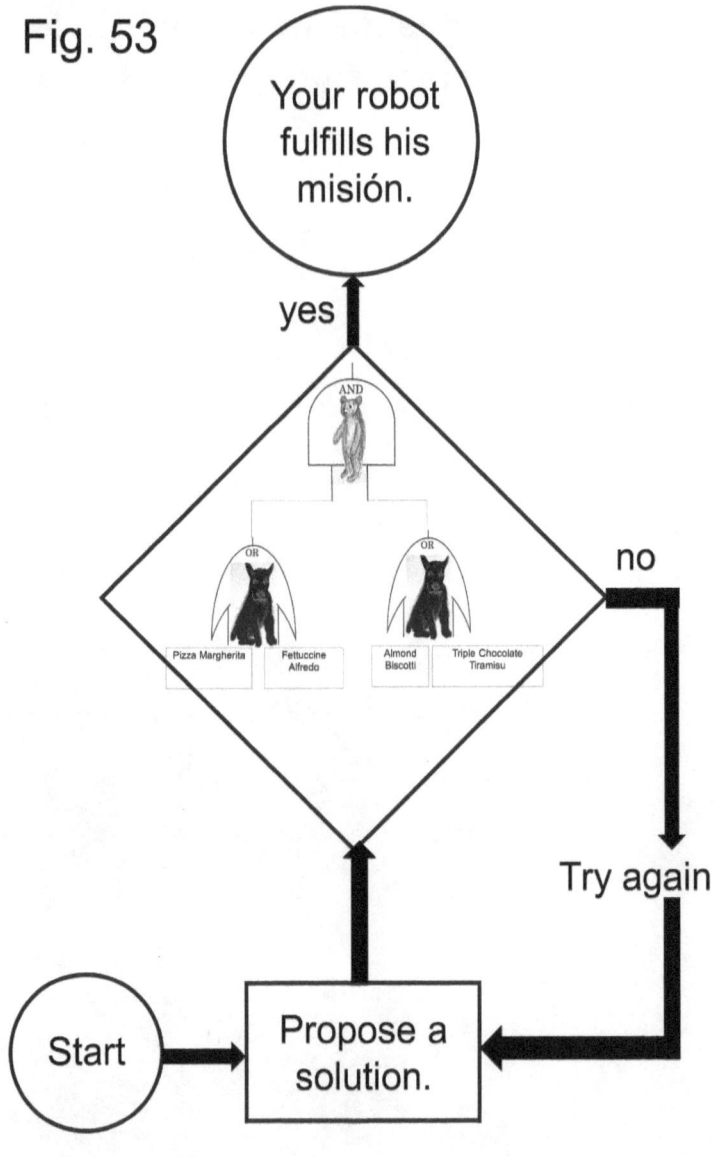

Please determine in the following list when your conditions are satisfied and your robot has found the correct menu.

Beef Bolognese with Linguine AND Triple Chocolate Tiramisu

Spaghetti with Roasted Red Pepper Sauce AND Chocolate Cannoli

Italian Baked Chicken AND Almond Biscotti

Pizza Margherita AND Strawberry Italian Ice

Garlic Tomato Bruschetta AND Triple Chocolate Tiramisu

Vegetable Lasagna AND Chocolate Cannoli

Fried Mozzarella Cheese Appetizers AND Almond Biscotti

Tuscan Pork Stew AND Strawberry Italian Ice

Traditional Lasagna AND Triple Chocolate Tiramisu

Tuscan Burgers with Pesto Mayo AND Chocolate Cannoli

Italian Spaghetti and Meatballs AND Almond Biscotti

Pasta Arrabbiata AND Strawberry Italian Ice

Cannelloni-Style Lasagna AND Triple Chocolate Tiramisu

Skillet Chicken with Olives AND Chocolate Cannoli

Butter Ravioli AND Almond Biscotti

Fettuccine alla Carbonara AND Strawberry Italian Ice Cream

Chicken with Marsala Risotto AND Triple Chocolate Tiramisu

Mushroom Ravioli Filling AND Chocolate Cannoli

Chicken Parmigiana AND Almond Biscotti

Italian Minestrone Soup AND Strawberry Italian Ice Cream

Panzanella AND Triple Chocolate Tiramisu

Mushroom Risotto with Peas AND Chocolate Cannoli

Grilled Polenta with Mushrooms AND Almond Biscotti

Meatball Pie AND Strawberry Italian Ice Cream

Fettuccine Alfredo AND Triple Chocolate Tiramisu

Prosciutto and Beans AND Chocolate Cannoli

Pesce Italiano AND Almond Biscotti

Will your dear robot deer be able to complete the task that you asked him to perform?

Fig. 54

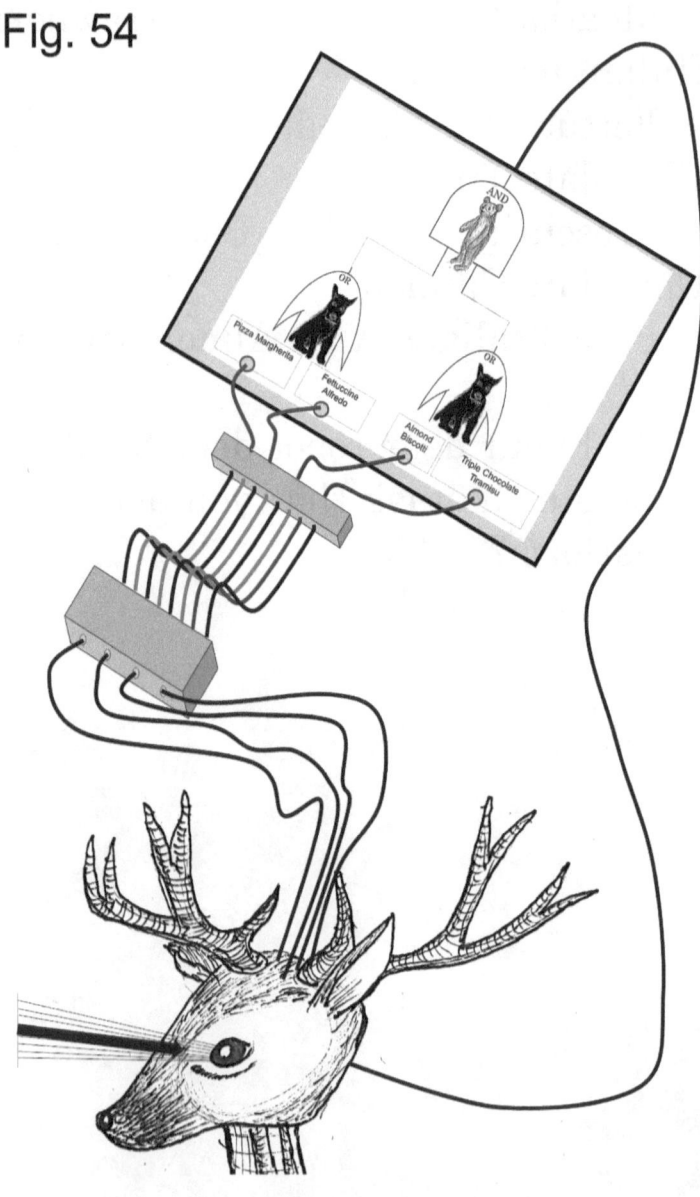